SJ Teacher's Choice Math Series

100 Challenging Algebra Problems & Solutions (Volume 2)

Assessment Lessons

For Grade 9 & 10 Students

SANJAY JAMINDAR

Copyright © 2017 Sanjay Jamindar

All rights reserved.

ISBN-10: 1545452946
ISBN-13: 978-1545452943

PREFACE

The aim of "**100 Challenging Algebra Problems & Solutions(Volume 2): Assessment Lessons For Grade 9 & 10 Students**" book is to help primary school students of Grade 9 and 10 (Class-IX, X) develop their Algebra problem solving skills and expand their knowledge of basic Algebra taught at Schools. The book provides ample practice on various types of problems which can be solved by basic Algebra Formulae. This is the second "Assessment Lessons" volume of the series of books to be published in future.

These problems will provide an overall assessment of the student's progress in learning basic Algebra concepts and formulae taught in various secondary class textbooks. Students will definitely find this book useful in preparing for their examinations and evaluating their knowledge of Algebra.

This book also provides the method of solving these problems along with the answers which are provided at the end of this book. Students are encouraged to consciously apply their original thoughts in solving these problems on their own.

I look forward to constructive criticism which will help me in improving this book.

Sanjay Jamindar

Bangalore, India

CONTENTS

1 PROBLEMS 6

2 SOLUTIONS AND ANSWERS 18

ALGEBRA BASIC FORMULAE

1. $(a+b)^2 = a^2 + 2ab + b^2$
2. $(a-b)^2 = a^2 - 2ab + b^2$
   $\Rightarrow (a+b)^2 = (a-b)^2 + 4ab$
   $\Rightarrow (a-b)^2 = (a+b)^2 - 4ab$
   $\Rightarrow a^2 + b^2 = (a+b)^2 - 2ab = (a-b)^2 + 2ab = \dfrac{1}{2}\left\{(a+b)^2 + (a-b)^2\right\}$
   $\Rightarrow ab = \left(\dfrac{a+b}{2}\right)^2 - \left(\dfrac{a-b}{2}\right)^2$

3. $(a+b)(a-b) = a^2 - b^2$

4. $(a+b)^3 = a^3 + 3a^2b + 3ab^2 + b^3$
   $\Rightarrow a^3 + b^3 = (a+b)^3 - 3ab(a+b)$

5. $(a-b)^3 = a^3 - 3a^2b + 3ab^2 - b^3$
   $\Rightarrow a^3 - b^3 = (a-b)^3 + 3ab(a-b)$

6. $a^3 + b^3 = (a+b)(a^2 - ab + b^2)$

7. $a^3 - b^3 = (a-b)(a^2 + ab + b^2)$

8. $(a+b+c)^2 = a^2 + b^2 + c^2 + 2(bc + ca + ab)$

9. $(a+b+c)^3 = a^3 + b^3 + c^3 + 3(a+b)(b+c)(c+a)$

10. $a^3 + b^3 + c^3 - 3abc = (a+b+c)(a^2 + b^2 + c^2 - ab - bc - ca)$
    $= \dfrac{1}{2}(a+b+c)\left\{(a-b)^2 + (b-c)^2 + (c-a)^2\right\}$
    $\Rightarrow a^2 + b^2 + c^2 - ab - bc - ca = \dfrac{1}{2}\left[(b-c)^2 + (c-a)^2 + (a-b)^2\right]$

PROBLEMS

1. If $x^3 + \dfrac{3}{x} = 4(a^3 + b^3)$ and $3x + \dfrac{1}{x^3} = 4(a^3 - b^3)$, prove that $a + b = \dfrac{1}{a-b}$.

2. Resolve into factors: $4x^8 + 81x^4 + 16x^4 y^4 + 324 y^4$.

3. If $ab + bc + ca = 0$, prove that $\dfrac{a}{a^2 - bc} + \dfrac{b}{b^2 - ca} + \dfrac{c}{c^2 - ab} = \dfrac{3}{a+b+c}$.

4. Resolve into factors: $(a+b-2c)^2 + a^2 - b^2 - 4c^2 + 4bc - a - b + 2c$.

5. If $\left(x + \dfrac{1}{x}\right)^2 = 3$, find the value of $x^{72} + x^{66} + x^{60} + x^{54} + x^{48} + x^{42} + x^{36} + x^{30} + x^{24} + x^{18} + x^{12} + x^6$.

6. Simplify: $\dfrac{a^6 - b^6}{a^4 + a^2 b^2 + b^4}$.

7. Simplify: $\dfrac{a^3 + 3a^2 + 18a + 54}{a^3 + 3a^2 + a + 3}$.

8. Raman can finish a work in 9 days and Rajan can finish the same work in 18 days. Both of them started the work together, but Raman left the job 3 days before the work is supposed to be finished. How many days it will take to finish the work?

9. Find the H.C.F of : $3x^2 - 11x - 4$ & $6x^3 - 25x^2 + 3$.

10. Solve: $\frac{1}{2}\left(x - \frac{a}{3}\right) - \frac{1}{3}\left(x - \frac{a}{4}\right) + \frac{1}{4}\left(x - \frac{a}{5}\right) = \frac{1}{5}\left(x - \frac{a}{6}\right)$.

11. Solve: $0.5x + \frac{0.02x + 0.07}{0.03} - \frac{x+2}{9} = 9.5$.

12. Arrange in ascending order: $\sqrt[4]{5}$, $\sqrt[8]{10}$, $\sqrt[12]{25}$, $\sqrt[6]{8}$.

13. Resolve into factors: $8a^6b + 27a^3b^4 - 8a^4b^3 - 27ab^6$.

14. Prove that: $\frac{1}{x^2 - 8x + 15} + \frac{1}{x^2 - 4x + 3} - \frac{2}{x^2 - 6x + 5} = \frac{x+a}{x-a} - \frac{x-a}{x+a} - \frac{4ax}{x^2 - a^2}$.

15. Simplify: $\frac{a}{a-x} + \frac{a}{a+x} + \frac{2a^2}{a^2 + x^2} + \frac{4a^4}{a^4 + x^4} - \frac{8a^8}{a^8 - x^8}$.

16. Solve: $\left(\frac{x+7}{x+14}\right)^2 = \frac{x+14}{x+28}$.

17. Solve: $\frac{x+2a}{b+c} + \frac{x+2b}{c+a} + \frac{x+2c}{a+b} = -6$.

18. There are two consecutive odd integers. $\left(\dfrac{1}{3}\right)^{rd}$ of the smaller number exceeds $\left(\dfrac{1}{5}\right)^{th}$ of the greater number by 4. Find the numbers.

19. If $x + \dfrac{1}{x} = 2$, find the value of $x^{177} - \dfrac{1}{x^{199}}$.

20. Resolve into factors: $3x^4 + 4x^2 - 4$.

21. If $4^{x+3} = 224 + 8 \times 4^x$, find the value of $(20x)^{2x+3}$.

22. Resolve into factors: $2a^2 - 4b^2 - 2c^2 + 6bc - 2ab$.

23. Resolve into factors: $2x^{10} + 2x^8 y^2 + 2x^6 y^4 - 16x^4 y^6 - 16x^2 y^8 - 16y^{10}$.

24. Show that $\dfrac{1}{9} + \dfrac{1}{25} + \dfrac{1}{49} + \dfrac{1}{81} \ldots\ldots + \dfrac{1}{(2k+1)^2} + \ldots\ldots + \dfrac{1}{2001^2} < \dfrac{1}{4}$.

25. If $x^2 - (a^2 + a - 2)y^2 + 3xy = 0$, find $x : y$.

26. If $x + \dfrac{16}{x} = 8$, then find the value of $x^4 + \dfrac{256}{x^4} - 257$.

27. Show that:
$\dfrac{(a-b)^2}{(b-c)(c-a)} + \dfrac{(b-c)^2}{(c-a)(a-b)} + \dfrac{(c-a)^2}{(a-b)(b-c)} = \dfrac{3a^2}{(a-b)(a-c)} + \dfrac{3b^2}{(b-c)(b-a)} + \dfrac{3c^2}{(c-a)(c-b)}$.

28. Find the L.C.M of: $4(a^4 - b^4)$, $2(a^4 + 2a^2b^2 + b^4)$, $6(a^4 - 3a^2b^2 - 4b^4)$.

29. If $x = 1 + \sqrt{2} + \sqrt{3}$ and $y = -1 + \sqrt{2} - \sqrt{3}$, show that $\dfrac{x^2 + 4xy + y^2}{x + y} = \sqrt{2}(1 - \sqrt{3})$.

30. Solve: $\dfrac{1}{2x+2} + \dfrac{3}{2x+3} + \dfrac{5}{2x+5} = \dfrac{9}{2x+4}$.

31. If $x = \dfrac{\sqrt{5}}{3}$, find the value of $\dfrac{\sqrt{1+x} - \sqrt{1-x}}{\sqrt{1+x} + \sqrt{1-x}}$.

32. Solve: $\dfrac{x^4 + 1}{x^2} = 289 \dfrac{1}{289}$.

33. If $a^2 + \dfrac{1}{a^2} = 398$, find the value of $a^3 + \dfrac{1}{a^3}$.

34. Solve: $\dfrac{2x+5}{x+1} = \dfrac{4x+5}{4x+4} + \dfrac{3x+3}{3x+1}$.

35. Resolve into factors: $2x^4 + 2x^3 - 4x^2 - 2x + 2$.

36. Resolve into factors: $20a^2 + 26ab - 6b^2$.

37. Solve: $\dfrac{4x+3}{2x+5} - \dfrac{3x-2}{3x+2} = 1$.

38. Show: $\dfrac{b+c}{4bc}(b^2+c^2-a^2)+\dfrac{c+a}{4ca}(c^2+a^2-b^2)+\dfrac{a+b}{4ab}(a^2+b^2-c^2)=\dfrac{1}{2}(a+b+c)$.

39. For a 3-digit integer number, the value of each digit is 1 less than its right side digit. If 27 is subtracted from the number, the result will be 16 times the sum of the digits. Find the number.

40. If $x+y=a$, $x^2+y^2=b^2$, $x^3+y^3=c^3$, show that $\dfrac{3a^3+6c^3}{b^2}=9a$.

41. If $3x-4y=5$, show that $27x^3-64y^3=125+180xy$.

42. Resolve into factors: $x^2-y^2-8xa+2ya+15a^2$.

43. Resolve into factors: $63x^3+72x^2-81x+54$.

44. Resolve into factors: x^8+68x^4+256.

45. Resolve into factors: $x^3+2x^2y-4xy^2-8y^3$.

46. In a fraction, the numerator is greater than the denominator. The difference between them is 7. If we add 1 to the numerator and multiply 2 with the denominator, the fraction becomes $\dfrac{1}{3}$. Find the fraction.

47. Solve: $\dfrac{2}{2x-3}+\dfrac{3}{3x+1}=\dfrac{6}{6x-1}+\dfrac{1}{x-1}$.

48. Solve $\left(\dfrac{x+10}{x-13}\right)^2 = \dfrac{(x+7)(x+13)}{(x-10)(x-16)}$.

49. At present, Peter's father is 3 times older than him. After 7 years, the sum of their ages would be 70. What is his father's current age?

50. If $x = \sqrt[3]{100} + 2$, prove that $x^3 - 6x^2 + 12x - 108 = 0$.

51. If $x^2 + y^2 + z^2 = 18$ & $xy + yz + zx = 9$, show that one of the values of $\dfrac{1}{1-x} + \dfrac{1}{1-y} + \dfrac{1}{1-z}$ is 0.

52. If $\dfrac{1}{a^2} + \dfrac{1}{b^2} + \dfrac{1}{c^2} = \dfrac{1}{ab} + \dfrac{1}{bc} + \dfrac{1}{ca}$, prove that $a = b = c$.

53. Resolve into factors: $x^2 - 20x + 51 - y^2 + 14y$.

54. Resolve into factors: $81x^5 + 4xy^4 - 81x^4y - 4y^5$.

55. If $a + 2b + 3c = 0$, prove that $\dfrac{10c}{a+c} - \dfrac{5a}{b+c} = 10$.

56. If $\dfrac{1}{a} + \dfrac{1}{b} + \dfrac{1}{c} = \dfrac{1}{a+b+c}$, prove that $\dfrac{3}{a^3} + \dfrac{3}{b^3} + \dfrac{3}{c^3} = \dfrac{3}{a^3+b^3+c^3} = \dfrac{3}{(a+b+c)^3}$.

57. The human population of a town was 20000. After three years, the male population increases by 10% and the female population decreases by 6%. But

the total population remains the same. What was the number of males and females, three years back?

58. If $\dfrac{1}{b+c} + \dfrac{1}{c+a} = \dfrac{2}{a+b}$, prove that $a^2 + b^2 = 2c^2$.

59. If $a + b = 1$, prove that $a^2 + b^2 + ab = a^3 + b^3 + 2ab$.

60. If $ab(a+b) = p^2$, prove that $a^3 + b^3 + 3p^2 = \dfrac{p^6}{a^3 b^3}$.

61. If $x = \sqrt{2} + \sqrt{3} - \sqrt{5}$ & $y = \sqrt{2} + \sqrt{3} + \sqrt{5}$, find the value of $\dfrac{x^2 + 18xy + y^2}{xy}$.

62. If $x^3 + \dfrac{3}{x} = 4(a^3 + b^3)$ & $3x + \dfrac{1}{x^3} = 4(a^3 - b^3)$, prove that $a^2 - b^2 - 1 = 0$.

63. If $(b + c - a)x = (c + a - b)y = (a + b - c)z = 3$,

prove that $\left(\dfrac{1}{y} + \dfrac{1}{z}\right)\left(\dfrac{1}{z} + \dfrac{1}{x}\right)\left(\dfrac{1}{x} + \dfrac{1}{y}\right) = \dfrac{8}{27} abc$.

64. Resolve into factors: $a^3 + ac - bd + bc - ad - ab^2$.

65. If $a + \dfrac{1}{b} = 1$ and $b + \dfrac{1}{c} = 1$, prove that $abc = -1$.

66. If $p = \dfrac{\sqrt{a^2+b^2}+\sqrt{a^2-b^2}}{\sqrt{a^2+b^2}-\sqrt{a^2-b^2}}$, prove that $b^2(p^2+1) = 2a^2 p$.

67. Solve: $2\left(\sqrt{\dfrac{x}{y}}+\sqrt{\dfrac{y}{x}}\right) = 5$, $x+y = 10$.

68. A person takes 12 seconds to walk down a moving escalator. If he walks down the stationary escalator at the same speed, he can come down in 21 seconds. In how many seconds will he be downstairs if he stands on a moving escalator?

69. Solve: $xy + x + y = 27$, $\dfrac{1}{x} + \dfrac{1}{y} = \dfrac{1}{8}$.

70. Solve: $\dfrac{(x+1)^3 - (x-1)^3}{(x+1)^2 - (x-1)^2} = 2$.

71. If $4ab = ax + bx$, show that $\dfrac{x+2a}{x-2a} + \dfrac{x+2b}{x-2b} = 2$.

72. If $a^2 - by - cz = 0$, $b^2 - cz - ax = 0$ and $c^2 - ax - by = 0$, prove that $\dfrac{x}{x+a} + \dfrac{y}{y+b} + \dfrac{z}{z+c} = 1$.

73. If $x = b-c$, $y = c-a$ and $z = a-b$, prove that $4x^2 - y^2 + 4z^2 + 8zx = 3(a-c)^2$.

74. If $x+y = \sqrt{5}$, $x-y = \sqrt{3}$, then find the value of $8xy(x^2+y^2)$.

75. Simplify: $\dfrac{x\left(\dfrac{a}{a-x}+\dfrac{b}{b-x}+\dfrac{c}{c-x}\right)}{\dfrac{3}{x}-\dfrac{1}{x-a}-\dfrac{1}{x-b}-\dfrac{1}{x-c}}$.

76. We divide 26 by 4 parts, such that if the first part is decreased by 3, the second part is increased by 11, the third part is multiplied by 4, and the fourth part is divided by 2, the results are all equal. Find all the parts.

77. Simplify: $\left(x^{16}-1\right)\left(\dfrac{1}{x-1}+\dfrac{1}{x+1}+\dfrac{2x}{x^2+1}+\dfrac{4x^3}{x^4+1}+\dfrac{8x^7}{x^8+1}\right)$.

78. Simplify: $\dfrac{b+c}{4bc}\left(b^2+c^2-a^2\right)+\dfrac{c+a}{4ca}\left(c^2+a^2-b^2\right)+\dfrac{a+b}{4ab}\left(a^2+b^2-c^2\right)$.

79. Solve: $(2x+1)+\dfrac{40}{(2x+1)}=41$.

80. Solve: $\dfrac{3x^2+4x+1}{6x^2+8x+5}=\dfrac{x^2+2x+7}{2x^2+4x+11}$.

81. If $x=3+2^{\frac{2}{3}}+2^{\frac{1}{3}}$, prove that $x^3-9x^2+21x+4=0$.

82. If $x=\dfrac{\sqrt{5}+1}{\sqrt{5}-1}$ and $y=\dfrac{\sqrt{5}-1}{\sqrt{5}+1}$, then find the value of $x^4+x^2y^2+y^4$.

83. If $x = \sqrt{\dfrac{p-1}{p+1}}$, prove that $\left(\dfrac{x}{x-1}\right)^2 + \left(\dfrac{x}{x+1}\right)^2 = p(p-1)$.

84. A container "A" has diluted milk where in every 5 liters of milk, 3 liters of water are mixed. In another container "B" the ratio of milk and water is $9:7$. A new solution of 30 liters is prepared in another container "C" after mixing certain amounts of diluted milk from the container "A" and "B" respectively. The amount of milk becomes $1\dfrac{1}{2}$ times the amount of water in the new container "C". How much diluted milk was taken out from each container to prepare the new solution?

85. If $\dfrac{a-b}{c} + \dfrac{b-c}{a} + \dfrac{c+a}{b} = 1$ & $a-b+c \neq 0$, prove that $a(b+c) = bc$.

86. Solve: $\dfrac{x+y}{2xy} = \dfrac{y+z}{2yz} = \dfrac{z+x}{2zx} = \dfrac{1}{3}$.

87. Solve: $ax + by - 1 = 0$, $bx + ay - \dfrac{(a+b)^2}{a^2+b^2} + 1 = 0$.

88. Find the time between 2 and 3 o'clock when the hands of a clock are together.

89. Prove that $\dfrac{\sqrt{10}-2}{5\sqrt{3}-\sqrt{32}-\sqrt{48}+\sqrt{18}} = \left(\sqrt{10}-2\right)\left(\sqrt{3}+\sqrt{2}\right)$.

90. If $\sqrt{5} = 3.342$, prove that $\dfrac{\sqrt{3+\sqrt{5}}}{\sqrt{2}-\sqrt{7-3\sqrt{5}}} = 1.85$.

91. Resolve into factors: $2x^2 + 4x + 2xy + 2y + 2$.

92. On a certain day, a car travelled at a constant speed and covered a certain distance in $2\frac{1}{2}$ hours. The next day, the car travelled at 2 km/hour higher speed than the previous day. However, the car covered 1 km lesser distance in the second day and the time taken to cover this distance was $\frac{1}{2}$ hour lesser than that of the previous day. What was the speed of the car on the first day?

93. A train crosses a 220 meter long bridge in 25 seconds and crosses another bridge of 120 meters long in 18 seconds. Find the length and speed of the train.

94. Prove that $0.7 + 0.\overline{7} + 0.\overline{57} = \frac{37}{18}$.

95. A cycle was sold at a profit of 25%. If the cost price was 10% less and selling price was $50 more, the shopkeeper could make a profit of 50%. What was the cost price of the cycle?

96. Solve: $\dfrac{1}{(x-1)(x-2)} + \dfrac{1}{(x-2)(x-3)} + \dfrac{1}{(x-3)(x-4)} = \dfrac{181}{1086}$.

97. Solve: $x^2 - x = 1482$.

98. Solve: $(x-2)(x-3) = \dfrac{34}{33^2}$.

99. If $x + y = 10$, find the largest value of xy. Assume, x and y are real numbers.

100. If $x^2 + y^2 + z^2 = 1$, prove that $-\frac{1}{2} \leq (xy + yz + zx) \leq 1$. Assume, x, y and z are integers.

SOLUTIONS AND ANSWERS

Problem 1:

$x^3 + \dfrac{3}{x} = 4(a^3 + b^3)$ .... (1)

$3x + \dfrac{1}{x^3} = 4(a^3 - b^3)$ .... (2)

Adding (1) and (2)

$x^3 + 3x + \dfrac{3}{x} + \dfrac{1}{x^3} = 8a^3$

$\Rightarrow \left(x + \dfrac{1}{x}\right)^3 = (2a)^3 \Rightarrow \left(x + \dfrac{1}{x}\right) = (2a)$ .... (3)

Subtracting (1) and (2)

$x^3 - 3x + \dfrac{3}{x} - \dfrac{1}{x^3} = 8b^3$

$\Rightarrow \left(x - \dfrac{1}{x}\right)^3 = (2b)^3 \Rightarrow \left(x - \dfrac{1}{x}\right) = (2b)$ .... (4)

Adding (3) and (4)

$2x = 2(a+b) \Rightarrow x = a+b$ .... (5)

Subtracting (3) and (4)

$\dfrac{2}{x} = 2(a-b) \Rightarrow \dfrac{1}{x} = a-b$ .... (6)

Multiplying (5) and (6)

$x \cdot \dfrac{1}{x} = (a+b)(a-b)$

$\Rightarrow a+b = \dfrac{1}{a-b}$ **(Ans.)**

Problem 2:

$4x^8 + 81x^4 + 16x^4 y^4 + 324 y^4$

$= 4x^8 + 16x^4 y^4 + 81x^4 + 324 y^4$ [Rearranging terms]

$$= 4x^4(x^4+4y^4)+81(x^4+4y^4)$$
$$= (x^4+4y^4)(4x^4+81)$$
$$= \{(x^2)^2+2.x^2.2y^2+(2y^2)^2-2.x^2.2y^2\}\{(2x^2)^2+2.2x^2.9+(9)^2-2.2x^2.9\}$$
$$= \{(x^2+2y^2)^2-(2xy)^2\}\{(2x^2+9)^2-(6x)^2\}$$
$$= (x^2+2y^2+2xy)(x^2+2y^2-2xy)(2x^2+9+6x)(2x^2+9-6x) \text{ (Ans.)}$$

Problem 3:

$\because ab+bc+ca=0$

$\Rightarrow ab+ca=-bc,\ bc+ab=-ca,\ bc+ca=-ab$

$$\therefore \frac{a}{a^2-bc}+\frac{b}{b^2-ca}+\frac{c}{c^2-ab}$$
$$= \frac{a}{a^2+ab+ca}+\frac{b}{b^2+bc+ab}+\frac{c}{c^2+bc+ca}$$
$$= \frac{a}{a(a+b+c)}+\frac{b}{b(a+b+c)}+\frac{c}{c(a+b+c)}$$
$$= \frac{3abc}{abc(a+b+c)}=\frac{3}{a+b+c} \text{ (Ans.)}$$

Problem 4:

$(a+b-2c)^2+a^2-b^2-4c^2+4bc-a-b+2c$
$= (a+b-2c)^2-(a+b-2c)+a^2-(b^2-4bc+4c^2)$ [Rearranging terms]
$= (a+b-2c)(a+b-2c-1)+a^2-(b-2c)^2$
$= (a+b-2c)(a+b-2c-1)+(a+b-2c)(a-b+2c)$
$= (a+b-2c)(a+b-2c-1+a-b+2c)$
$= (a+b-2c)(2a-1)$ **(Ans.)**

Problem 5:

$$\left(x+\frac{1}{x}\right)^2=3$$

$\Rightarrow x^2+\dfrac{1}{x^2}=1 \Rightarrow x^4+1=x^2 \Rightarrow x^4-x^2=-1$

$x^{72} + x^{66} + x^{60} + x^{54} + x^{48} + x^{42} + x^{36} + x^{30} + x^{24} + x^{18} + x^{12} + x^6$
$= x^{66}(x^6 + 1) + x^{54}(x^6 + 1) + x^{42}(x^6 + 1) + x^{30}(x^6 + 1) + x^{18}(x^6 + 1) + x^6(x^6 + 1)$
$= (x^6 + 1)(x^{66} + x^{54} + x^{42} + x^{30} + x^{18} + x^6)$

Now $x^6 + 1$
$= (x^2)^3 + 1^3$
$= (x^2 + 1)(x^4 - x^2 + 1) = (x^2 + 1)(-1 + 1) = 0 \quad [\because x^4 - x^2 = -1]$

$\therefore x^{72} + x^{66} + x^{60} + x^{54} + x^{48} + x^{42} + x^{36} + x^{30} + x^{24} + x^{18} + x^{12} + x^6$
$= (x^6 + 1)(x^{66} + x^{54} + x^{42} + x^{30} + x^{18} + x^6) = 0$ **(Ans.)**
$[\because x^6 + 1 = 0]$

Problem 6:
$\dfrac{a^6 - b^6}{a^4 + a^2 b^2 + b^4}$

$= \dfrac{(a^3)^2 - (b^3)^2}{(a^2 + b^2)^2 - (ab)^2}$

$= \dfrac{(a^3 + b^3)(a^3 - b^3)}{(a^2 + b^2 + ab)(a^2 + b^2 - ab)}$

$= \dfrac{(a+b)(a^2 - ab + b^2)(a-b)(a^2 + ab + b^2)}{(a^2 + b^2 + ab)(a^2 + b^2 - ab)} = (a+b)(a-b)$ **(Ans.)**

$\because a^3 + b^3 = (a+b)(a^2 - ab + b^2), a^3 - b^3 = (a-b)(a^2 + ab + b^2)]$

Problem 7:
$a^3 + 3a^2 + 18a + 54$
$= a^3 + 27 + 3a^2 + 18a + 27$ [Rearranging and splitting terms]
$= (a^3 + 3^3) + 3(a^2 + 6a + 9)$
$= (a+3)(a^2 - 3a + 9) + 3(a+3)^2$
$= (a+3)(a^2 + 18)$

$a^3 + 3a^2 + a + 3$
$= a^2(a+3) + (a+3)$
$= (a^2+1)(a+3)$

$\therefore \dfrac{a^3 + 3a^2 + 18a + 54}{a^3 + 3a^2 + a + 3} = \dfrac{(a+3)(a^2+18)}{(a^2+1)(a+3)} = \dfrac{(a^2+18)}{(a^2+1)}$ **(Ans.)**

Problem 8:

Let x is the number of days it took to finish the work.

According to the question, Raman worked for $(x-3)$ days and Rajan worked for x days.

$\therefore$ Raman finished $\dfrac{x-3}{9}$ part and Rajan finished $\dfrac{x}{18}$ part of the work.

$\therefore \dfrac{x-3}{9} + \dfrac{x}{18} = 1 \Rightarrow x = 8$

It will take 8 days to finish the work. **(Ans.)**

Problem 9:

$3x^2 - 11x - 4$
$= 3x^2 - 12x + x - 4$
$= 3x(x-4) + 1(x-4)$
$= (x-4)(3x+1)$

$6x^3 - 25x^2 + 3$
$= 6x^3 + 2x^2 - 27x^2 - 9x + 9x + 3$
$= 2x^2(3x+1) - 9x(3x+1) + 3(3x+1)$
$= (3x+1)(2x^2 - 9x + 3)$

The H.C.F $= (3x+1)$ **(Ans.)**

Problem 10:

$\dfrac{1}{2}\left(x - \dfrac{a}{3}\right) - \dfrac{1}{3}\left(x - \dfrac{a}{4}\right) + \dfrac{1}{4}\left(x - \dfrac{a}{5}\right) = \dfrac{1}{5}\left(x - \dfrac{a}{6}\right)$

$\Rightarrow \dfrac{3x-a}{6} - \dfrac{4x-a}{12} + \dfrac{5x-a}{20} = \dfrac{6x-a}{30}$

$\Rightarrow \dfrac{30x - 10a - 20x + 5a + 15x - 3a}{60} = \dfrac{6x-a}{30}$

$\Rightarrow \dfrac{25x - 8a}{60} = \dfrac{6x-a}{30}$

$\Rightarrow 25x - 8a = 12x - 2a$

$\Rightarrow x = \dfrac{6a}{13}$ **(Ans.)**

Problem 11:

$0.5x + \dfrac{0.02x + 0.07}{0.03} - \dfrac{x+2}{9} = 9.5$

$\Rightarrow \dfrac{x}{2} + \dfrac{\dfrac{x}{50} + \dfrac{7}{100}}{\dfrac{3}{100}} - \dfrac{x+2}{9} = \dfrac{19}{2}$

$\Rightarrow \dfrac{x}{2} + \dfrac{\dfrac{2x+7}{100}}{\dfrac{3}{100}} - \dfrac{x+2}{9} = \dfrac{19}{2}$

$\Rightarrow \dfrac{x}{2} + \dfrac{2x+7}{3} - \dfrac{x+2}{9} = \dfrac{19}{2}$

$\Rightarrow \dfrac{57x + 126 - 12}{54} = \dfrac{19}{2}$

$\Rightarrow 57x + 114 = 513$

$\Rightarrow x = 7$ **(Ans.)**

Problem 12:

$\sqrt[4]{5} = 5^{\frac{1}{4}} = 5^{\frac{6}{24}} = \left(5^6\right)^{\frac{1}{24}} = 15625^{\frac{1}{24}}$

$\sqrt[8]{10} = 10^{\frac{1}{8}} = 10^{\frac{3}{24}} = \left(10^3\right)^{\frac{1}{24}} = 1000^{\frac{1}{24}}$

$\sqrt[12]{25} = 25^{\frac{1}{12}} = 25^{\frac{2}{24}} = \left(25^2\right)^{\frac{1}{24}} = 625^{\frac{1}{24}}$

$$\sqrt[6]{8} = 8^{\frac{1}{6}} = 8^{\frac{4}{24}} = \left(8^4\right)^{\frac{1}{24}} = 4096^{\frac{1}{24}}$$

$$\therefore \sqrt[12]{25} < \sqrt[8]{10} < \sqrt[6]{8} < \sqrt[4]{5} \quad \textbf{(Ans.)}$$

Problem 13:

$8a^6b + 27a^3b^4 - 8a^4b^3 - 27ab^6$
$= 8a^6b - 8a^4b^3 + 27a^3b^4 - 27ab^6 \quad \text{[Rearranging terms]}$
$= 8a^4b(a^2 - b^2) + 27ab^4(a^2 - b^2)$
$= (a^2 - b^2)(8a^4b + 27ab^4)$
$= ab(a^2 - b^2)(8a^3 + 27b^3)$
$= ab(a^2 - b^2)(2a + 3b)(4a^2 - 6ab + 9b^2) \quad [\because a^3 + b^3 = (a+b)(a^2 - ab + b^2)]$
$= ab(a+b)(a-b)(2a+3b)(4a^2 - 6ab + 9b^2) \quad \textbf{(Ans.)}$

Problem 14:

$$\frac{1}{x^2 - 8x + 15} + \frac{1}{x^2 - 4x + 3} - \frac{2}{x^2 - 6x + 5}$$

$$= \frac{1}{(x-3)(x-5)} + \frac{1}{(x-1)(x-3)} - \frac{2}{(x-1)(x-5)}$$

$$= \frac{(x-1) + (x-5) - 2(x-3)}{(x-1)(x-3)(x-5)}$$

$$= \frac{x - 1 + x - 5 - 2x + 6}{(x-1)(x-3)(x-5)} = 0$$

$$\frac{x+a}{x-a} - \frac{x-a}{x+a} - \frac{4ax}{x^2 - a^2}$$

$$= \frac{(x+a)^2 - (x-a)^2 - 4ax}{x^2 - a^2} = 0$$

$$\therefore \frac{1}{x^2 - 8x + 15} + \frac{1}{x^2 - 4x + 3} - \frac{2}{x^2 - 6x + 5} = \frac{x+a}{x-a} - \frac{x-a}{x+a} - \frac{4ax}{x^2 - a^2} = 0 \quad \textbf{(Ans.)}$$

Problem 15:

$$\frac{a}{a-x}+\frac{a}{a+x}+\frac{2a^2}{a^2+x^2}+\frac{4a^4}{a^4+x^4}-\frac{8a^8}{a^8-x^8}$$

$$=\frac{2a^2}{a^2-x^2}+\frac{2a^2}{a^2+x^2}+\frac{4a^4}{a^4+x^4}-\frac{8a^8}{a^8-x^8}$$

$$=\frac{4a^4}{a^4-x^4}+\frac{4a^4}{a^4+x^4}-\frac{8a^8}{a^8-x^8}$$

$$=\frac{8a^8}{a^8-x^8}-\frac{8a^8}{a^8-x^8}=0 \textbf{ (Ans.)}$$

Problem 16:

$$\left(\frac{x+7}{x+14}\right)^2=\frac{x+14}{x+28}$$

$$\Rightarrow \frac{(x+7)^2}{x+14}=\frac{(x+14)^2}{x+28}$$

$$\Rightarrow \frac{x^2+14x+49}{x+14}=\frac{x^2+28x+196}{x+28}$$

$$\Rightarrow \frac{x^2+14x}{x+14}+\frac{49}{x+14}=\frac{x^2+28x}{x+28}+\frac{196}{x+28}$$

$$\Rightarrow x+\frac{49}{x+14}=x+\frac{196}{x+28}$$

$$\Rightarrow \frac{49}{x+14}=\frac{196}{x+28}$$

$$\Rightarrow 49(x+28)=196(x+14)$$

$$\Rightarrow x=-\frac{1372}{147} \textbf{ (Ans.)}$$

Problem 17:

$$\frac{x+2a}{b+c}+\frac{x+2b}{c+a}+\frac{x+2c}{a+b}=-6$$

$$\Rightarrow \left(\frac{x+2a}{b+c}+2\right)+\left(\frac{x+2b}{c+a}+2\right)+\left(\frac{x+2c}{a+b}+2\right)=-6+6 \text{ [Adding 6 in both sides]}$$

$$\Rightarrow \left(\frac{x+2a+2b+2c}{b+c}\right)+\left(\frac{x+2b+2c+2a}{c+a}\right)+\left(\frac{x+2c+2a+2b}{a+b}\right)=0$$

$$\Rightarrow (x+2a+2b+2c)\left(\frac{1}{b+c}+\frac{1}{c+a}+\frac{1}{a+b}\right)=0$$

$$\Rightarrow (x+2a+2b+2c)=0 \quad \left[\text{As }\left(\frac{1}{b+c}+\frac{1}{c+a}+\frac{1}{a+b}\right)\text{ cannot be 0}\right]$$

$$\Rightarrow x=-(2a+2b+2c) \textbf{ (Ans.)}$$

Problem 18:

Let x is an integer. Therefore, $2x+1$ is one of the odd integer and the next odd integer is $2x+3$.

According to the question:

$$\frac{2x+1}{3}-\frac{2x+3}{5}=8$$

$$\Rightarrow \frac{10x+5-6x-9}{15}=8$$

$$\Rightarrow 4x-4=120 \Rightarrow x=31$$

$\therefore$ The two numbers are $2x+1=63$ & $2x+3=65$. **(Ans.)**

Problem 19:

$x+\dfrac{1}{x}=2 \Rightarrow (x-1)^2=0 \Rightarrow x=1$

$\therefore x^{177}-\dfrac{1}{x^{199}}=1-1=0$ **(Ans.)**

Problem 20:

$3x^4+4x^2-4$

$= 2x^4+x^4+4x^2-8+4$ [Rearranging terms]

$= (2x^4-8)+(x^4+4x^2+4)$ [Rearranging terms]

$= 2(x^4-4)+(x^2+2)^2$

$= 2(x^2+2)(x^2-2)+(x^2+2)^2$

$= (x^2+2)(2x^2-4+x^2+2)$

$= (x^2 + 2)(3x^2 - 2)$ **(Ans.)**

Problem 21:
$4^{x+3} = 224 + 8 \times 4^x$

$\Rightarrow (2^2)^{x+3} = 224 + 2^3 \times (2^2)^x$

$\Rightarrow 2^{2x+6} = 224 + 2^{2x+3}$

$\Rightarrow 2^{2x+6} - 2^{2x+3} = 224$

$\Rightarrow 2^{2x}(2^6 - 2^3) = 224$

$\Rightarrow 2^{2x} \times 56 = 224$

$\Rightarrow 2^{2x} = 2^2$

$\Rightarrow x = 1$

$\therefore (20x)^{2x+3} = (20)^5 = 32 \times 10^5$ **(Ans.)**

Problem 22:
$2a^2 - 4b^2 - 2c^2 + 6bc - 2ab$

$= a^2 + a^2 - 4b^2 - c^2 - c^2 + 4bc + 2bc - 2ab + b^2 - b^2$ [Splitting terms]

$= (a^2 - 2ab + b^2) - (4b^2 - 4bc + c^2) + a^2 - (b^2 - 2bc + c^2)$ [Rearranging terms]

$= \{(a-b)^2 - (2b-c)^2\} + \{a^2 - (b-c)^2\}$

$= (a+b-c)(a-3b+c) + (a+b-c)(a-b+c)$

$= (a+b-c)\{(a-3b+c) + (a-b+c)\}$

$= (a+b-c)(2a-4b+2c)$

$= 2(a+b-c)(a-2b+c)$ **(Ans.)**

Problem 23:
$2x^{10} + 2x^8 y^2 + 2x^6 y^4 - 16x^4 y^6 - 16x^2 y^8 - 16y^{10}$

$= 2x^4(x^6 - 8y^6) + 2x^2 y^2(x^6 - 8y^6) + 2y^4(x^6 - 8y^6)$

$= 2(x^6 - 8y^6)(x^4 + x^2 y^2 + y^4)$

$= 2(x^2 - 2y^2)(x^4 + 2x^2 y^2 + 4y^4)(x^2 + xy + y^2)(x^2 - xy + y^2)$ **(Ans.)**

Problem 24:

$$\frac{1}{(2k+1)^2} < \frac{1}{(2k+1)^2 - 1}$$

$$\Rightarrow \frac{1}{(2k+1)^2} < \frac{1}{2k(2k+1)}$$

$$\Rightarrow \frac{1}{(2k+1)^2} < \left(\frac{1}{4}\right)\left(\frac{1}{k} - \frac{1}{k+1}\right)$$

$$\therefore \frac{1}{9} + \frac{1}{25} + \frac{1}{49} + \frac{1}{81} \ldots\ldots + \frac{1}{(2k+1)^2} + \ldots\ldots + \frac{1}{2001^2}$$

$$< \frac{1}{4}\left(\frac{1}{1} - \frac{1}{2}\right) + \frac{1}{4}\left(\frac{1}{2} - \frac{1}{3}\right) + \ldots\ldots + \frac{1}{4}\left(\frac{1}{1000} - \frac{1}{1001}\right)$$

$$= \frac{1}{4}\left(1 - \frac{1}{1001}\right)$$

$$\therefore \frac{1}{9} + \frac{1}{25} + \frac{1}{49} + \frac{1}{81} \ldots\ldots + \frac{1}{(2k+1)^2} + \ldots\ldots + \frac{1}{2001^2} < \frac{1}{4} \textbf{ (Ans.)}$$

Problem 25:

$x^2 - (a^2 + a - 2)y^2 + 3xy = 0$

$\Rightarrow x^2 + 3xy = (a^2 + a - 2)y^2$

$\Rightarrow \dfrac{x^2 + 3xy}{y^2} = (a^2 + a - 2)$

$\Rightarrow \dfrac{x^2}{y^2} + \dfrac{3x}{y} = (a^2 + a - 2)$

$\Rightarrow m^2 + 3m = (a^2 + a - 2)$; Assume, $m = \dfrac{x}{y}$

$\Rightarrow m^2 + 3m - (a^2 + a - 2) = 0$

$\Rightarrow m^2 + 3m - (a+2)(a-1) = 0$

$\Rightarrow m^2 + \{(a+2) - (a-1)\}m - (a+2)(a-1) = 0$

$\Rightarrow m^2 + (a+2)m - (a-1)m - (a+2)(a-1) = 0$

$\Rightarrow m(m + a + 2) - (a-1)(m + a + 2) = 0$

$\Rightarrow (m + a + 2)(m - a + 1) = 0$

∴ Either $(m + a + 2) = 0$

$$\Rightarrow m = -(a+2) \Rightarrow \frac{x}{y} = -(a+2)$$

Or, $(m-a+1) = 0$

$$\Rightarrow m = (a-1) \Rightarrow \frac{x}{y} = (a-1)$$

$\therefore x:y = -(a+2)$ or $x:y = (a-1)$ **(Ans.)**

Problem 26:

$$x + \frac{16}{x} = 8$$
$$\Rightarrow x^2 - 8x + 16 = 0$$
$$\Rightarrow (x-4)^2 = 0 \Rightarrow x = 4$$
$$\therefore x^4 + \frac{256}{x^4} - 257 = 256 + 1 - 257 = 0 \text{ \textbf{(Ans.)}}$$

Problem 27:

L.H.S=

$$\frac{(a-b)^2}{(b-c)(c-a)} + \frac{(b-c)^2}{(c-a)(a-b)} + \frac{(c-a)^2}{(a-b)(b-c)}$$

$$= \frac{(a-b)^3 + (b-c)^3 + (c-a)^3}{(a-b)(b-c)(c-a)}$$

$$= \frac{3(a-b)(b-c)(c-a)}{(a-b)(b-c)(c-a)} = 3 \quad [\because (a-b)+(b-c)+(c-a)=0]$$

R.H.S=

$$\frac{3a^2}{(a-b)(a-c)} + \frac{3b^2}{(b-c)(b-a)} + \frac{3c^2}{(c-a)(c-b)}$$

$$= -3\left\{\frac{a^2}{(a-b)(c-a)} + \frac{b^2}{(b-c)(a-b)} + \frac{c^2}{(c-a)(b-c)}\right\}$$

$$= -3\left\{\frac{a^2(b-c) + b^2(c-a) + c^2(a-b)}{(a-b)(c-a)(b-c)}\right\}$$

Now $a^2(b-c)+b^2(c-a)+c^2(a-b)$
$= a^2b - a^2c + b^2c - ab^2 + ac^2 - bc^2 + abc - abc$
$= a^2b - a^2c + b^2c - ab^2 + ac^2 - bc^2 + abc - abc$

$= (a^2b - abc) - (ab^2 - b^2c) + (abc - bc^2) - (a^2c - ac^2)$
$= ab(a-c) - b^2(a-c) + bc(a-c) - ac(a-c)$
$= (a-c)(ab - b^2 + bc - ac)$
$= (a-c)(ab - b^2 + bc - ac)$
$= (a-c)\{b(a-b) - c(a-b)\}$
$= -(a-b)(c-a)(b-c)$

$\therefore$ R.H.S $= -3\left\{\dfrac{-(a-b)(c-a)(b-c)}{(a-b)(c-a)(b-c)}\right\} = 3$

$\therefore$ L.H.S=R.H.S **(Ans.)**

Problem 28:

$4(a^4 - b^4) = 4(a^2 + b^2)(a+b)(a-b)$
$2(a^4 + 2a^2b^2 + b^4) = 2(a^2 + b^2)^2$
$6(a^4 - 3a^2b^2 - 4b^4) = 6(a^2 + b^2)(a^2 - 4b^2) = 6(a^2 + b^2)(a+2b)(a-2b)$
$\therefore$ L.C.M $= 12(a^2 + b^2)^2(a^2 - b^2)(a^2 - 4b^2)$ **(Ans.)**

Problem 29:

Given that $x = 1 + \sqrt{2} + \sqrt{3}$ and $y = -1 + \sqrt{2} - \sqrt{3}$
$\therefore x + y = 2\sqrt{2}$ and $xy = -2(1 + \sqrt{3})$
$\therefore \dfrac{x^2 + 4xy + y^2}{x+y}$
$= \dfrac{(x+y)^2 + 2xy}{x+y}$
$= \dfrac{8 - 4(1+\sqrt{3})}{2\sqrt{2}} = \dfrac{4 - 4\sqrt{3}}{2\sqrt{2}} = \dfrac{2(1-\sqrt{3})}{\sqrt{2}} = \sqrt{2}(1-\sqrt{3})$ **(Ans.)**

Problem 30:

$$\frac{1}{2x+2} + \frac{3}{2x+3} + \frac{5}{2x+5} = \frac{9}{2x+4}$$

$$\Rightarrow \frac{1}{2x+2} + \frac{3}{2x+3} + \frac{5}{2x+5} = \frac{1}{2x+4} + \frac{3}{2x+4} + \frac{5}{2x+4}$$

$$\Rightarrow \left(\frac{1}{2x+2} - \frac{1}{2x+4}\right) + \left(\frac{3}{2x+3} - \frac{3}{2x+4}\right) = \left(\frac{5}{2x+4} - \frac{5}{2x+5}\right)$$

$$\Rightarrow \frac{2}{(2x+2)(2x+4)} + \frac{3}{(2x+3)(2x+4)} = \frac{5}{(2x+4)(2x+5)}$$

$$\Rightarrow \frac{2}{(2x+2)} + \frac{3}{(2x+3)} = \frac{5}{(2x+5)}$$

$$\Rightarrow \frac{10x+12}{(2x+2)(2x+3)} = \frac{5}{(2x+5)}$$

$$\Rightarrow 20x^2 + 74x + 60 = 20x^2 + 50x + 30$$

$$\Rightarrow 24x = -30 \Rightarrow x = -\frac{5}{4} \quad \textbf{(Ans.)}$$

Problem 31:

$$\frac{\sqrt{1+x} - \sqrt{1-x}}{\sqrt{1+x} + \sqrt{1-x}}$$

$$= \frac{(\sqrt{1+x} - \sqrt{1-x})(\sqrt{1+x} - \sqrt{1-x})}{(\sqrt{1+x} + \sqrt{1-x})(\sqrt{1+x} - \sqrt{1-x})}$$

$$= \frac{(\sqrt{1+x} - \sqrt{1-x})^2}{(\sqrt{1+x})^2 - (\sqrt{1-x})^2}$$

$$= \frac{(1+x+1-x-2\sqrt{(1+x)(1-x)})}{1+x-1+x}$$

$$= \frac{(2 - 2\sqrt{(1+x)(1-x)})}{2x} = \frac{1-\sqrt{1-x^2}}{x} = \frac{1}{\sqrt{5}} \quad \textbf{(Ans.)}$$

$$[\because x = \frac{\sqrt{5}}{3}]$$

Problem 32:

$$\frac{x^4+1}{x^2} = 289\frac{1}{289}$$

$$\Rightarrow x^2 + \frac{1}{x^2} = 289 + \frac{1}{289}$$

$$\Rightarrow x^2 - 289 = \frac{1}{289} - \frac{1}{x^2}$$

$$\Rightarrow x^2 - 289 = \frac{x^2 - 289}{289x^2}$$

$$\Rightarrow (x^2 - 289) - \frac{(x^2 - 289)}{289x^2} = 0$$

$$\Rightarrow (x^2 - 289)\left(1 - \frac{1}{289x^2}\right) = 0$$

$\therefore$ Either $(x^2 - 289) = 0 \Rightarrow x^2 = 289 \Rightarrow x = \pm 17$

Or $\left(1 - \frac{1}{289x^2}\right) = 0 \Rightarrow x^2 = \frac{1}{289} \Rightarrow x = \pm\frac{1}{17}$

$\therefore x = 17, -17, \frac{1}{17}, -\frac{1}{17}$ **(Ans.)**

Problem 33:

$$a^2 + \frac{1}{a^2} = 398$$

$$\Rightarrow \left(a + \frac{1}{a}\right)^2 - 2 = 398 \Rightarrow a + \frac{1}{a} = \pm 20$$

$$\therefore a^3 + \frac{1}{a^3} = \left(a + \frac{1}{a}\right)^3 - 3\left(a + \frac{1}{a}\right) = \pm 7940 \text{ **(Ans.)**}$$

Problem 34:

$$\frac{2x+5}{x+1} = \frac{4x+5}{4x+4} + \frac{3x+3}{3x+1}$$

$$\Rightarrow \left(\frac{2x+5}{x+1} - 2\right) = \left(\frac{4x+5}{4x+4} - 1\right) + \left(\frac{3x+3}{3x+1} - 1\right)$$

$$\Rightarrow \left(\frac{2x+5-2x-2}{x+1}\right) = \left(\frac{4x+5-4x-4}{4x+4}\right) + \left(\frac{3x+3-3x-1}{3x+1}\right)$$

$$\Rightarrow \left(\frac{3}{x+1}\right) = \left(\frac{1}{4x+4}\right) + \left(\frac{2}{3x+1}\right)$$

$$\Rightarrow \left(\frac{3}{x+1}\right) - \left(\frac{1}{4x+4}\right) = \left(\frac{2}{3x+1}\right)$$

$$\Rightarrow \frac{11}{4(x+1)} = \frac{2}{3x+1}$$

$$\Rightarrow 33x + 11 = 8x + 8$$

$$\Rightarrow x = -\frac{3}{25} \text{ (Ans.)}$$

Problem 35:

$$2x^4 + 2x^3 - 4x^2 - 2x + 2$$
$$= 2(x^4 + x^3 - 2x^2 - x + 1)$$
$$= 2\{(x^4 - 2x^2 + 1) + (x^3 - x)\}$$
$$= 2\{(x^2 - 1)^2 + x(x^2 - 1)\}$$
$$= 2(x^2 - 1)(x^2 + x - 1)$$
$$= 2(x+1)(x-1)(x^2 + x - 1) \text{ (Ans.)}$$

Problem 36:

$$20a^2 + 26ab - 6b^2$$
$$= 2(10a^2 + 13ab - 3b^2)$$
$$= 2(10a^2 + 15ab - 2ab - 3b^2)$$
$$= 2\{5a(2a+3b) - b(2a+3b)\}$$
$$= 2(2a+3b)(5a-b) \text{ (Ans.)}$$

Problem 37:

$$\frac{4x+3}{2x+5} - \frac{3x-2}{3x+2} = 1$$

$$\Rightarrow \frac{(4x+10)-7}{2x+5} - \frac{(3x+2)-4}{3x+2} = 1$$

$$\Rightarrow 2 - \frac{7}{2x+5} - 1 + \frac{4}{3x+2} = 1$$

$$\Rightarrow \frac{4}{3x+2} = \frac{7}{2x+5}$$

$$\Rightarrow 4(2x+5) = 7(3x+2)$$

$$\Rightarrow x = \frac{6}{13} \text{ (Ans.)}$$

Problem 38:

$$\frac{b+c}{4bc} = \frac{b}{4bc} + \frac{c}{4bc} = \frac{1}{4c} + \frac{1}{4b}$$

$$\frac{c+a}{4ca} = \frac{c}{4ca} + \frac{a}{4ca} = \frac{1}{4a} + \frac{1}{4c}$$

$$\frac{a+b}{4ab} = \frac{a}{4ab} + \frac{b}{4ab} = \frac{1}{4b} + \frac{1}{4a}$$

$$\therefore \frac{b+c}{4bc}(b^2+c^2-a^2) + \frac{c+a}{4ca}(c^2+a^2-b^2) + \frac{a+b}{4ab}(a^2+b^2-c^2)$$

$$= \left(\frac{1}{4c} + \frac{1}{4b}\right)(b^2+c^2-a^2) + \left(\frac{1}{4a} + \frac{1}{4c}\right)(c^2+a^2-b^2) + \left(\frac{1}{4b} + \frac{1}{4a}\right)(a^2+b^2-c^2)$$

$$= \frac{1}{4a}(c^2+a^2-b^2+a^2+b^2-c^2) + \frac{1}{4b}(b^2+c^2-a^2+a^2+b^2-c^2) +$$

$$\frac{1}{4c}(c^2+a^2-b^2+b^2+c^2-a^2)$$

$$= \frac{1}{4a}(2a^2) + \frac{1}{4b}(2b^2) + \frac{1}{4c}(2c^2)$$

$$= \frac{1}{2}(a+b+c) \text{ (Ans.)}$$

Problem 39:

If the digit in the hundreds place is x, the digit in the tens place is $x+1$ and the digit in the units place is $x+2$.

∴ The number is $\{100x + 10(x+1) + (x+2)\} = 111x + 12$

According to the question,

$(111x + 12) - 27 = 16\{x + (x+1) + (x+2)\}$

$\Rightarrow 111x - 15 = 16(3x + 3)$

$\Rightarrow x = 1$

∴ The number is $= 111 \times 1 + 12 = 123$ **(Ans.)**

Problem 40:

$\dfrac{3a^3 + 6c^3}{b^2}$

$= \dfrac{3(x+y)^3 + 6(x^3 + y^3)}{(x^2 + y^2)}$

$= \dfrac{3\{(x^3 + y^3) + 3xy(x+y)\} + 6(x^3 + y^3)}{(x^2 + y^2)}$

$= \dfrac{9xy(x+y) + 9(x^3 + y^3)}{(x^2 + y^2)}$

$= \dfrac{9xy(x+y) + 9(x+y)(x^2 - xy + y^2)}{(x^2 + y^2)}$

$= \dfrac{9(x+y)(x^2 + y^2)}{(x^2 + y^2)} = 9(x+y) = 9a$ **(Ans.)**

$[\because x + y = a]$

Problem 41:

$3x - 4y = 5$

$\Rightarrow (3x - 4y)^3 = 125$

$\Rightarrow 27x^3 - 108x^2y + 144xy^2 - 64y^3 = 125$

$\Rightarrow 27x^3 - 64y^3 - 36xy(3x - 4y) = 125$

$\Rightarrow 27x^3 - 64y^3 = 125 + 36xy(3x - 4y)$

$\Rightarrow 27x^3 - 64y^3 = 125 + 180xy$ **(Ans.)**

Problem 42:

$x^2 - y^2 - 8xa + 2ya + 15a^2$
$= [x^2 - 2.x.4a + (4a)^2] - [a^2 - 2ya + y^2]$
$= (x - 4a)^2 - (a - y)^2$
$= (x - 4a + a - y)(x - 4a - a + y)$
$= (x - 3a - y)(x - 5a + y)$ **(Ans.)**

Problem 43:

$63x^3 + 72x^2 - 81x + 54$
$= 9(7x^3 + 8x^2 - 9x + 6)$
$= 9(7x^3 + 14x^2 - 6x^2 - 12x + 3x + 6)$
$= 9[7x^2(x + 2) - 6x(x + 2) + 3(x + 2)]$
$= 9(x + 2)(7x^2 - 6x + 3)$ **(Ans.)**

Problem 44:

$x^8 + 68x^4 + 256$
$= x^8 + 4x^4 + 64x^4 + 256$
$= x^4(x^4 + 4) + 64(x^4 + 4)$
$= (x^4 + 4)(x^4 + 64)$
$= [(x^2 + 2)^2 - (2x)^2][(x^2 + 8)^2 - (4x)^2]$
$= (x^2 + 2x + 2)(x^2 - 2x + 2)(x^2 + 4x + 8)(x^2 - 4x + 8)$ **(Ans.)**

Problem 45:

$x^3 + 2x^2y - 4xy^2 - 8y^3$
$= [x^3 - (2y)^3] + 2xy(x - 2y)$
$= (x - 2y)(x^2 + 2xy + 4y^2) + 2xy(x - 2y)$
$= (x - 2y)[x^2 + 4xy + 4y^2]$
$= (x - 2y)(x + 2y)(x + 2y)$ **(Ans.)**

Problem 46:

According to the question, the fraction is $\dfrac{x}{x+7}$, where x = numerator.

According to the second condition,

$\dfrac{x+1}{2(x+7)} = \dfrac{1}{3} \Rightarrow x = 11$

$\therefore$ The required fraction is $\dfrac{11}{18}$ **(Ans.)**

Problem 47:

$\dfrac{2}{2x-3} + \dfrac{3}{3x+1} = \dfrac{6}{6x-1} + \dfrac{1}{x-1}$

$\Rightarrow \dfrac{2}{2x-3} - \dfrac{1}{x-1} = \dfrac{6}{6x-1} - \dfrac{3}{3x+1}$

$\Rightarrow \dfrac{1}{2x^2 - 5x + 3} = \dfrac{9}{18x^2 + 3x - 1}$

$\Rightarrow 18x^2 + 3x - 1 = 18x^2 - 45x + 27$

$\therefore x = \dfrac{7}{12}$ **(Ans.)**

Problem 48:

$\left(\dfrac{x+10}{x-13}\right)^2 = \dfrac{(x+7)(x+13)}{(x-10)(x-16)}$

$\Rightarrow \dfrac{x^2 + 20x + 100}{x^2 - 26x + 169} = \dfrac{x^2 + 20x + 91}{x^2 - 26x + 160}$

$\Rightarrow \dfrac{x^2 + 20x + 100}{x^2 + 20x + 91} = \dfrac{x^2 - 26x + 169}{x^2 - 26x + 160}$

$\Rightarrow \dfrac{x^2 + 20x + 100}{x^2 + 20x + 91} - 1 = \dfrac{x^2 - 26x + 169}{x^2 - 26x + 160} - 1$

$\Rightarrow \dfrac{x^2 + 20x + 100 - x^2 - 20x - 91}{x^2 + 20x + 84} = \dfrac{x^2 - 26x + 169 - x^2 + 26x - 160}{x^2 - 26x + 160}$

$\Rightarrow \dfrac{9}{x^2 + 20x + 84} = \dfrac{9}{x^2 - 26x + 165}$

$\Rightarrow x^2 + 20x + 84 = x^2 - 26x + 165$

$\Rightarrow 46x = 81$

$\therefore x = \dfrac{81}{46}$ **(Ans.)**

Problem 49:

According to the question, if Peter's current age is x, his father's current age would be $3x$.

After 7 years, their ages would be $(x+7)$ and $(3x+7)$ respectively.

$\therefore (x+7) + (3x+7) = 70$

$\Rightarrow x = 14$

$\therefore$ His father's current age is 42 years. **(Ans.)**

Problem 50:

$x^3 - 6x^2 + 12x - 108$

$= x^3 - 3.x^2.2 + 3.x.2^2 - 2^3 - 100$

$= (x-2)^3 - 100 = 0$ **(Ans.)**

$[\because x = \sqrt[3]{100} + 2 \Rightarrow (x-2)^3 = 100]$

Problem 51:

$x^2 + y^2 + z^2 = (x+y+z)^2 - 2(xy+yz+zx)$

$\therefore (x+y+z)^2 = x^2+y^2+z^2 + 2(xy+yz+zx) = 18+18 = 36$

$\therefore (x+y+z) = \pm 6$

$\therefore \dfrac{1}{1-x} + \dfrac{1}{1-y} + \dfrac{1}{1-z}$

$= \dfrac{3 - 2(x+y+z) + (xy+yz+zx)}{(1-x)(1-y)(1-z)}$

$= \dfrac{3 - 2.6 + 9}{(1-x)(1-y)(1-z)} = 0$ **(Ans.)**

$[\because (x+y+z) = +6]$

Problem 52:

$\dfrac{1}{a^2} + \dfrac{1}{b^2} + \dfrac{1}{c^2} = \dfrac{1}{ab} + \dfrac{1}{bc} + \dfrac{1}{ca}$

$\Rightarrow \dfrac{b^2c^2 + c^2a^2 + a^2b^2}{a^2b^2c^2} = \dfrac{a+b+c}{abc}$

$\Rightarrow b^2c^2 + c^2a^2 + a^2b^2 = abc(a+b+c)$

$\Rightarrow b^2c^2 + c^2a^2 + a^2b^2 - a^2bc - ab^2c - abc^2 = 0$

$\Rightarrow x^2 + y^2 + z^2 - xy - yz - zx = 0$ [Assume $x = bc$, $y = ca$ and $z = ab$]

$\Rightarrow \dfrac{1}{2}\{(x-y)^2 + (y-z)^2 + (z-x)^2\} = 0$

If the sum of square terms is zero, each must be equal to zero.

$\Rightarrow x - y = y - z = z - x = 0$

$\Rightarrow a = b = c$ **(Ans.)**

Problem 53:

$x^2 - 20x + 51 - y^2 + 14y$

$= (x^2 - 2.x.10 + 100) - (y^2 - 2.y.7 + 49)$

$= (x-10)^2 - (y-7)^2$

$= (x - 10 + y - 7)(x - 10 - y + 7)$

$= (x + y - 17)(x - y - 3)$ **(Ans.)**

Problem 54:

$81x^5 + 4xy^4 - 81x^4y - 4y^5$

$= 81x^5 - 81x^4y + 4xy^4 - 4y^5$

$= 81x^4(x - y) + 4y^4(x - y)$

$= (x - y)(81x^4 + 4y^4)$

$= (x - y)\{(9x^2)^2 + (2y^2)^2\}$

$= (x - y)(9x^2 + 2y^2 + 6xy)(9x^2 + 2y^2 - 6xy)$ **(Ans.)**

Problem 55:

$a + 2b + 3c = 0$

$\Rightarrow a+c = -2(b+c)$

$\therefore \dfrac{10c}{a+c} - \dfrac{5a}{b+c}$

$= \dfrac{10c}{-2(b+c)} - \dfrac{5a}{b+c}$

$= \dfrac{-5(c+a)}{(b+c)} = 10$ **(Ans.)**

Problem 56:

$\dfrac{1}{a} + \dfrac{1}{b} + \dfrac{1}{c} = \dfrac{1}{a+b+c}$

$\Rightarrow (a+b+c)(bc+ca+ab) = abc$
$\Rightarrow (a+b+c)(bc+ca+ab) - abc = 0$
$\Rightarrow (b+c)(c+a)(a+b) = 0$

One of the above term must be zero, since the product is zero.

$\therefore a+b = 0 \Rightarrow a = -b \Rightarrow a^3 = -b^3$

$\dfrac{3}{a^3} + \dfrac{3}{b^3} + \dfrac{3}{c^3} = \dfrac{3}{a^3+b^3+c^3} = \dfrac{3}{(a+b+c)^3} = \dfrac{3}{c^3}$ **(Ans.)**

Problem 57:

Suppose, the male population was x and the female population was y.

From the 1st condition, $x + y = 20000$ ....(1)

From the 2nd condition, $\dfrac{110x}{100} + \dfrac{94y}{100} = 20000$ ....(2)

Solving (1) and (2), we get $x = 7500$ and $y = 12500$.

The male population was 7500 and the female population was 12500. **(Ans.)**

Problem 58:

$\dfrac{1}{b+c} + \dfrac{1}{c+a} = \dfrac{2}{a+b}$

$\Rightarrow \dfrac{1}{b+c} + \dfrac{1}{c+a} = \dfrac{1}{a+b} + \dfrac{1}{a+b}$

$\Rightarrow \dfrac{1}{b+c} - \dfrac{1}{a+b} = \dfrac{1}{a+b} - \dfrac{1}{c+a}$

$\Rightarrow (a-c)(c+a) = (c-b)(b+c)$

$\Rightarrow a^2 + b^2 = 2c^2$ **(Ans.)**

Problem 59:

R.H.S $= a^3 + b^3 + 2ab$

$= (a+b)(a^2 - ab + b^2) + 2ab$

$= a^2 + ab + b^2 = $ L.H.S **(Ans.)**

Problem 60:

$a^3 + b^3 + 3p^2$

$= (a+b)^3 - 3ab(a+b) + 3p^2$

$= \dfrac{p^6}{a^3 b^3} - 3ab \cdot \dfrac{p^2}{ab} + 3p^2$

$= \dfrac{p^6}{a^3 b^3}$ **(Ans.)**

Problem 61:

$x + y = 2(\sqrt{3} + \sqrt{2})$

$xy = 2\sqrt{6}$

$\therefore x^2 + 18xy + y^2$

$= (x+y)^2 + 16xy$

$= 4(5 + 2\sqrt{6}) + 32\sqrt{6}$

$= 20 + 40\sqrt{6}$

$= 20(1 + 2\sqrt{6})$

$\therefore \dfrac{x^2 + 18xy + y^2}{xy} = \dfrac{20(1 + 2\sqrt{6})}{2\sqrt{6}} = 10\left(2 + \dfrac{1}{\sqrt{6}}\right)$ **(Ans.)**

Problem 62:

$$x^3 + \frac{3}{x} = 4(a^3 + b^3) \quad \ldots(1)$$

$$3x + \frac{1}{x^3} = 4(a^3 - b^3) \quad \ldots(2)$$

Adding (1) and (2),

$$x^3 + \frac{1}{x^3} + \frac{3}{x} + 3x = 8a^3 \implies x + \frac{1}{x} = 2a \quad \ldots(3)$$

Subtracting (1) and (2),

$$x^3 - \frac{1}{x^3} + \frac{3}{x} - 3x = 8b^3 \implies x - \frac{1}{x} = 2b \quad \ldots(4)$$

Adding (3) and (4),

$$a + b = x \quad \ldots(5)$$

Subtracting (3) and (4),

$$a - b = \frac{1}{x} \quad \ldots(6)$$

From (5) and (6),

$$a^2 - b^2 = (a+b)(a-b) = x \cdot \frac{1}{x} = 1$$

$$\implies a^2 - b^2 - 1 = 0 \quad \textbf{(Ans.)}$$

Problem 63:

$$(b+c-a)x = (c+a-b)y = (a+b-c)z = 3$$

$$\therefore \frac{1}{x} = \frac{b+c-a}{3}, \frac{1}{y} = \frac{c+a-b}{3}, \frac{1}{z} = \frac{a+b-c}{3}$$

$$\therefore \left(\frac{1}{y} + \frac{1}{z}\right)\left(\frac{1}{z} + \frac{1}{x}\right)\left(\frac{1}{x} + \frac{1}{y}\right) = \frac{1}{3} \cdot 2a \cdot \frac{1}{3} \cdot 2b \cdot \frac{1}{3} \cdot 2c = \frac{8}{27}abc \quad \textbf{(Ans.)}$$

Problem 64:

$$a^3 + ac - bd + bc - ad - ab^2$$

$$= a^3 - ab^2 + ac - bd + bc - ad \quad \text{[Rearranging terms]}$$

$$= a(a^2 - b^2) + (ac + bc) - (bd + ad)$$

$$= a(a+b)(a-b) + c(a+b) - d(b+a)$$
$$= a(a+b)(a-b) + (a+b)(c-d)$$
$$= (a+b)(a^2 - ab + c - d) \text{ (Ans.)}$$

Problem 65:

$$a + \frac{1}{b} = 1 \Rightarrow a = \frac{b-1}{b}$$

$$b + \frac{1}{c} = 1 \Rightarrow c = \frac{1}{1-b}$$

$$\therefore 2abc + 1 = 2 \cdot \frac{b-1}{b} \cdot b \cdot \frac{1}{1-b} + 1 = -1$$

$$\Rightarrow abc = -1 \text{ (Ans.)}$$

Problem 66:

$$p = \frac{\sqrt{a^2 + b^2} + \sqrt{a^2 - b^2}}{\sqrt{a^2 + b^2} - \sqrt{a^2 - b^2}}$$

$$\Rightarrow \frac{p+1}{p-1} = \frac{\sqrt{a^2+b^2} + \sqrt{a^2-b^2} + \sqrt{a^2+b^2} - \sqrt{a^2-b^2}}{\sqrt{a^2+b^2} + \sqrt{a^2-b^2} - \sqrt{a^2+b^2} + \sqrt{a^2-b^2}}$$

$$\Rightarrow \frac{p+1}{p-1} = \frac{\sqrt{a^2+b^2}}{\sqrt{a^2-b^2}}$$

$$\Rightarrow \left(\frac{p+1}{p-1}\right)^2 = \left(\frac{\sqrt{a^2+b^2}}{\sqrt{a^2-b^2}}\right)^2$$

$$\Rightarrow \frac{(p+1)^2}{(p-1)^2} = \frac{a^2+b^2}{a^2-b^2}$$

$$\Rightarrow \frac{(p+1)^2 + (p-1)^2}{(p+1)^2 - (p-1)^2} = \frac{a^2+b^2+a^2-b^2}{a^2+b^2-a^2+b^2}$$

$$\Rightarrow \frac{2(p^2+1)}{2.2p} = \frac{2a^2}{2b^2}$$

$$\Rightarrow \frac{(p^2+1)}{2p} = \frac{a^2}{b^2}$$

$$\Rightarrow b^2(p^2+1) = 2a^2 p \text{ (Ans.)}$$

Problem 67:

$$\sqrt{\frac{x}{y}} + \sqrt{\frac{y}{x}} = \frac{5}{2}$$

$$\Rightarrow \frac{x+y}{\sqrt{xy}} = \frac{5}{2}$$

$$\Rightarrow \frac{10}{\sqrt{xy}} = \frac{5}{2}$$

$$\Rightarrow xy = 16 \Rightarrow x = \frac{16}{y}$$

From the eq. (2), $\frac{16}{y} + y = 10$

$$\Rightarrow y^2 - 10y + 16 = 0$$
$$\Rightarrow (y-8)(y-2) = 0$$
$$\Rightarrow y = 8 \text{ OR } y = 2$$
$$\therefore x = 2 \text{ OR } x = 8$$

The solutions are

$$\begin{cases} x = 2 \\ y = 8 \end{cases} \text{ OR } \begin{cases} x = 8 \\ y = 2 \end{cases} \text{ (Ans.)}$$

Problem 68:

Assume the speed of the escalator = x meter/s and the speed of the person = y meter/s.

∴ The length of the escalator = $21y$ meter.

When the person walks down the moving escalator, the combined speed with which he comes down = $(x+y)$ meter/s and the distance covered by him = $12(x+y)$ meter.

Since the distance covered is equal to the length of the escalator,

∴ $12(x+y) = 21y$

∴ $x = \frac{3y}{4}$

Now the time for the escalator to move through a distance of $21y$ meter = $\dfrac{21y}{x}$ sec.

∴ The person can be downstairs if he stands on a moving escalator for

$$\dfrac{21y}{x} = \dfrac{21y \times 4}{3y} = 28 \text{ sec.} \textbf{ (Ans.)}$$

Problem 69:

$xy + x + y = 27$ ....(1)

$\dfrac{1}{x} + \dfrac{1}{y} = \dfrac{1}{2}$ ....(2)

From (2), $xy = 2(x + y)$ ....(3)

From (3) and (1):

$2(x + y) + (x + y) = 27$

$\Rightarrow 3(x + y) = 27 \Rightarrow x + y = 9$ ....(4)

From (3) and (4):

$xy = 18$

Now, $(x - y)^2 = (x + y)^2 - 4xy = 81 - 4.18 = 9$

∴ $x - y = \sqrt{9} = \pm 3$ ....(5)

∴ $\begin{cases} x + y = 9 \\ x - y = 3 \end{cases}$ ....(6)   OR $\begin{cases} x + y = 9 \\ x - y = -3 \end{cases}$ ....(7)

From (6) we get, $x = 6$, $y = 3$

From (7) we get, $x = 3$, $y = 6$

The solutions are:

$\begin{cases} x = 6 \\ y = 3 \end{cases}$ OR $\begin{cases} x = 3 \\ y = 6 \end{cases}$ **(Ans.)**

Problem 70:

$\dfrac{(x+1)^3 - (x-1)^3}{(x+1)^2 - (x-1)^2} = 2$

$$\Rightarrow \frac{(x^3+3x^2+3x+1)-(x^3-3x^2+3x-1)}{(x^2+2x+1)-(x^2-2x+1)}=2$$

$$\Rightarrow \frac{6x^2+2}{4x}=2$$

$$\Rightarrow 3x^2-4x+1=0 \Rightarrow (x-1)(3x-1)=0$$

$$\therefore x=1 \text{ OR } x=\frac{1}{3} \textbf{ (Ans.)}$$

Problem 71:

$$\frac{x+2a}{x-2a}+\frac{x+2b}{x-2b}$$

$$=\left(\frac{x+2a}{x-2a}-1\right)+\left(\frac{x+2b}{x-2b}-1\right)+2$$

$$=\left(\frac{4a}{x-2a}\right)+\left(\frac{4b}{x-2b}\right)+2$$

$$=\frac{4(ax-2ab+bx-2ab)}{(x-2a)(x-2b)}+2$$

$$=\frac{4\{(a+b)x-4ab\}}{(x-2a)(x-2b)}+2$$

$$=\frac{4\times 0}{(x-2a)(x-2b)}+2=2 \textbf{ (Ans.)}$$

$[\because 4ab=ax+bx]$

Problem 72:

$$\frac{x}{x+a}+\frac{y}{y+b}+\frac{z}{z+c}$$

$$=\frac{ax}{ax+a^2}+\frac{by}{by+b^2}+\frac{cz}{cz+c^2}$$

$$=\frac{ax}{ax+by+cz}+\frac{by}{ax+by+cz}+\frac{cz}{ax+by+cz}$$

$[\because a^2-by-cz=0, \ b^2-cz-ax=0 \text{ and } c^2-ax-by=0]$

$=1$ **(Ans.)**

Problem 73:
$4x^2 - y^2 + 4z^2 + 8zx$
$= 4x^2 + 4z^2 + 8zx - y^2$
$= (2x + 2z)^2 - y^2$
$= (2x + y + 2z)(2x - y + 2z)$
$= 3(a-c)^2$ **(Ans.)**
$[\because x = b-c, \ y = c-a \text{ and } z = a-b]$

Problem 74:
$x + y = \sqrt{5}$ ....(1)
$x - y = \sqrt{3}$ ....(2)
Adding (1) and (2)
$x = \dfrac{\sqrt{5} + \sqrt{3}}{2}$

Subtracting (1) and (2)
$y = \dfrac{\sqrt{5} - \sqrt{3}}{2}$
$\therefore xy = 1$
$\therefore 8xy(x^2 + y^2)$
$= 8xy\{(x-y)^2 + 2xy\}$
$= 40$ **(Ans.)**

Problem 75:
The denominator=
$\dfrac{3}{x} - \dfrac{1}{x-a} - \dfrac{1}{x-b} - \dfrac{1}{x-c}$
$= \left(\dfrac{1}{x} + \dfrac{1}{a-x}\right) + \left(\dfrac{1}{x} + \dfrac{1}{b-x}\right) + \left(\dfrac{1}{x} + \dfrac{1}{c-x}\right)$
$= \left(\dfrac{a}{x(a-x)}\right) + \left(\dfrac{b}{x(b-x)}\right) + \left(\dfrac{c}{x(c-x)}\right)$

$$= \frac{1}{x}\left(\frac{a}{a-x}+\frac{b}{b-x}+\frac{c}{c-x}\right)$$

$$\therefore \frac{x\left(\frac{a}{a-x}+\frac{b}{b-x}+\frac{c}{c-x}\right)}{\frac{3}{x}-\frac{1}{x-a}-\frac{1}{x-b}-\frac{1}{x-c}}$$

$$= \frac{x\left(\frac{a}{a-x}+\frac{b}{b-x}+\frac{c}{c-x}\right)}{\frac{1}{x}\left(\frac{a}{a-x}+\frac{b}{b-x}+\frac{c}{c-x}\right)} = x^2 \textbf{ (Ans.)}$$

Problem 76:

Let the 4 parts are a, b, c and d.

Then $a+b+c+d = 26$ .... (1)

And $a-3 = b+11 = 4c = \dfrac{d}{2}$ .... (2)

From (2) let's assume, $a-3 = b+11 = 4c = \dfrac{d}{2} = k$

$\therefore a = k+3$, $b = k-11$, $c = \dfrac{k}{4}$, $d = 2k$ .... (3)

Putting (3) into (1) we get,

$(k+3)+(k-11)+\left(\dfrac{k}{4}\right)+2k = 26$

$\Rightarrow 4(k+3)+4(k-11)+4\left(\dfrac{k}{4}\right)+4.2k = 104$

$\Rightarrow 17k = 104-12+44 = 136$

$\therefore k = 8$

$\therefore a = 11$, $b = -3$, $c = 2$, $d = 16$ **(Ans.)**

Problem 77:

$$\left(x^{16}-1\right)\left(\frac{1}{x-1}+\frac{1}{x+1}+\frac{2x}{x^2+1}+\frac{4x^3}{x^4+1}+\frac{8x^7}{x^8+1}\right)$$

$$=\left(x^{16}-1\right)\left(\frac{2x}{x^2-1}+\frac{2x}{x^2+1}+\frac{4x^3}{x^4+1}+\frac{8x^7}{x^8+1}\right)$$

$$=\left(x^{16}-1\right)\left(\frac{4x^3}{x^4-1}+\frac{4x^3}{x^4+1}+\frac{8x^7}{x^8+1}\right)$$

$$=\left(x^{16}-1\right)\left(\frac{8x^7}{x^8-1}+\frac{8x^7}{x^8+1}\right)$$

$$=\left(x^{16}-1\right)\frac{16x^{15}}{\left(x^{16}-1\right)}$$

$$=16x^{15} \quad \textbf{(Ans.)}$$

Problem 78:

$$\frac{b+c}{4bc}\left(b^2+c^2-a^2\right)+\frac{c+a}{4ca}\left(c^2+a^2-b^2\right)+\frac{a+b}{4ab}\left(a^2+b^2-c^2\right)$$

$$=\left(\frac{1}{4c}+\frac{1}{4b}\right)\left(b^2+c^2-a^2\right)+\left(\frac{1}{4a}+\frac{1}{4c}\right)\left(c^2+a^2-b^2\right)+\left(\frac{1}{4b}+\frac{1}{4a}\right)\left(a^2+b^2-c^2\right)$$

$$=\frac{1}{4a}\left(a^2+b^2-c^2+c^2+a^2-b^2\right)+\frac{1}{4b}\left(a^2+b^2-c^2+b^2+c^2-a^2\right)$$
$$+\frac{1}{4c}\left(b^2+c^2-a^2+c^2+a^2-b^2\right)$$

$$=\frac{1}{4a}\left(2a^2\right)+\frac{1}{4b}\left(2b^2\right)+\frac{1}{4c}\left(2c^2\right)$$

$$=\frac{1}{2}(a+b+c) \quad \textbf{(Ans.)}$$

Problem 79:

$$(2x+1)+\frac{40}{(2x+1)}=41$$

$$\Rightarrow (2x+1)+\frac{40}{(2x+1)}=1+40$$

$\Rightarrow (2x+1)-1 = 40 - \dfrac{40}{(2x+1)}$

$\Rightarrow 2x = 40\left(1 - \dfrac{1}{(2x+1)}\right)$

$\Rightarrow 2x = \dfrac{40 \times 2x}{2x+1}$

$\Rightarrow 2x+1 = 40 \Rightarrow x = \dfrac{39}{2}$ **(Ans.)**

Problem 80:

$\dfrac{3x^2+4x+4}{6x^2+8x+5} = \dfrac{x^2+2x+6}{2x^2+4x+11}$

$\Rightarrow 2\left(\dfrac{3x^2+4x+4}{6x^2+8x+5}\right) = 2\left(\dfrac{x^2+2x+6}{2x^2+4x+11}\right)$

$\Rightarrow \left(\dfrac{6x^2+8x+8}{6x^2+8x+5}\right) = \left(\dfrac{2x^2+4x+12}{2x^2+4x+11}\right)$

$\Rightarrow \dfrac{(6x^2+8x+5)+3}{6x^2+8x+5} = \dfrac{(2x^2+4x+11)+1}{2x^2+4x+11}$

$\Rightarrow 1 + \dfrac{3}{6x^2+8x+5} = 1 + \dfrac{1}{2x^2+4x+11}$

$\Rightarrow \dfrac{3}{6x^2+8x+5} = \dfrac{1}{2x^2+4x+11}$

$\Rightarrow 3(2x^2+4x+11) = (6x^2+8x+5)$

$\Rightarrow 6x^2+12x+33 = 6x^2+8x+5$

$\Rightarrow 4x = -28$

$\therefore x = -7$ **(Ans.)**

Problem 81:

$x = 3 + 2^{\frac{2}{3}} + 2^{\frac{1}{3}}$

$\Rightarrow (x-3)^3 = \left(2^{\frac{2}{3}} + 2^{\frac{1}{3}}\right)^3$

$\Rightarrow x^3 - 9x^2 + 27x - 8 = \left(2^{\frac{2}{3}}\right)^3 + \left(2^{\frac{1}{3}}\right)^3 + 3.2^{\frac{2}{3}}.2^{\frac{1}{3}}\left(2^{\frac{2}{3}} + 2^{\frac{1}{3}}\right)$

$\Rightarrow x^3 - 9x^2 + 27x - 8 = 6 + 6(x-3)$

$\Rightarrow x^3 - 9x^2 + 27x - 8 = 6x - 12$

$\Rightarrow x^3 - 9x^2 + 21x + 4 = 0$ **(Ans.)**

Problem 82:

$x^4 + x^2y^2 + y^4$

$= x^4 + 2x^2y^2 + y^4 - x^2y^2$

$= \left(x^2 + y^2\right)^2 - (xy)^2$

$= \left(x^2 + xy + y^2\right)\left(x^2 - xy + y^2\right)$

Now, $x + y = \dfrac{\sqrt{5}+1}{\sqrt{5}-1} + \dfrac{\sqrt{5}-1}{\sqrt{5}+1}$

$= \dfrac{\left(\sqrt{5}+1\right)^2 + \left(\sqrt{5}-1\right)^2}{4}$

$= \dfrac{6 + 2\sqrt{5} + 6 - 2\sqrt{5}}{4}$

$= \dfrac{12}{4} = 3$

And, $xy = \dfrac{\sqrt{5}+1}{\sqrt{5}-1} \cdot \dfrac{\sqrt{5}-1}{\sqrt{5}+1} = 1$

$\therefore x^2 + xy + y^2$

$= (x+y)^2 - xy$

$= 9 - 1 = 8$

$x^2 - xy + y^2$

$= (x+y)^2 - 3xy$

$= 9 - 3 = 6$

$$\therefore x^4 + x^2y^2 + y^4$$
$$= (x^2 + xy + y^2)(x^2 - xy + y^2)$$
$$= 8 \times 6 = 48 \textbf{ (Ans.)}$$

Problem 83:

$$\left(\frac{x}{x-1}\right)^2 + \left(\frac{x}{x+1}\right)^2$$

$$= \frac{x^2}{(x-1)^2} + \frac{x^2}{(x+1)^2}$$

$$= x^2 \times \frac{2(x^2+1)}{(x^2-1)^2}$$

$$= \frac{p-1}{p+1} \times 2 \left\{ \left(\frac{p-1}{p+1} + 1\right) \div \left(\frac{p-1}{p+1} - 1\right)^2 \right\}$$

$$= \frac{p-1}{p+1} \times 2 \times \frac{2p}{p+1} \times \frac{(p+1)^2}{4} = p(p-1) \textbf{ (Ans.)}$$

Problem 84:

Let x liters of solutions are taken out from the container "A" and $(30-x)$ liters of solutions are taken out from the container "B".

There will be $\frac{5x}{8}$ liters of milk and $\frac{3x}{8}$ liters of water in the x liters of solution of the container "A".

There will be $\frac{9}{16}(30-x)$ liters of milk and $\frac{7}{16}(30-x)$ liters of water in the $(30-x)$ liters of solution of the container "B".

$\therefore$ In the new solution of the container "C",

the amount of milk = $\frac{5x}{8} + \frac{9}{16}(30-x)$ liters.

And the amount of water = $\frac{3x}{8} + \frac{7}{16}(30-x)$ liters.

According to the question,

$$\frac{5x}{8} + \frac{9}{16}(30-x) = 1\frac{1}{2}\left\{\frac{3x}{8} + \frac{7}{16}(30-x)\right\}$$

$\Rightarrow x = 18 \quad \therefore (30-x) = 12$

$\therefore$ 18 liters of solutions are taken out from the container "A" and 12 liters of solutions are taken out from the container "B". **(Ans.)**

Problem 85:

$\dfrac{a-b}{c} + \dfrac{b-c}{a} + \dfrac{c+a}{b} = 1$

$\Rightarrow \dfrac{a-b}{c} + \dfrac{b-c}{a} + \dfrac{c+a}{b} - 1 = 0$

$\Rightarrow \dfrac{a-b}{c} + 1 + \dfrac{b-c}{a} - 1 + \dfrac{c+a}{b} - 1 = 0$

$\Rightarrow \dfrac{a-b+c}{c} - \dfrac{a-b+c}{a} + \dfrac{a-b+c}{b} = 0$

$\Rightarrow (a-b+c)\left(\dfrac{1}{c} - \dfrac{1}{a} + \dfrac{1}{b}\right) = 0$

$\therefore \left(\dfrac{1}{c} - \dfrac{1}{a} + \dfrac{1}{b}\right) = 0$ Since $(a-b+c) \neq 0$

$\Rightarrow \dfrac{1}{b} + \dfrac{1}{c} = \dfrac{1}{a}$

$\Rightarrow a(b+c) = bc$ **(Ans.)**

Problem 86:

$\dfrac{x+y}{xy} = \dfrac{2}{3} \quad \ldots (1)$

$\dfrac{y+z}{yz} = \dfrac{2}{3} \quad \ldots (2)$

$\dfrac{z+x}{zx} = \dfrac{2}{3} \quad \ldots (3)$

From (1), $\dfrac{1}{y}+\dfrac{1}{x}=\dfrac{2}{3}$ .... (4)

From (2), $\dfrac{1}{z}+\dfrac{1}{y}=\dfrac{2}{3}$ .... (5)

From (3), $\dfrac{1}{x}+\dfrac{1}{z}=\dfrac{2}{3}$ .... (6)

Adding (4), (5) and (6)

$2\left(\dfrac{1}{x}+\dfrac{1}{y}+\dfrac{1}{z}\right)=3\cdot\dfrac{2}{3}$

$\Rightarrow \left(\dfrac{1}{x}+\dfrac{1}{y}+\dfrac{1}{z}\right)=1$ .... (7)

Subtracting (7) with (4),

$\left(\dfrac{1}{x}+\dfrac{1}{y}+\dfrac{1}{z}\right)-\left(\dfrac{1}{y}+\dfrac{1}{x}\right)=1-\dfrac{2}{3}$

$\Rightarrow \dfrac{1}{z}=\dfrac{1}{3} \quad \Rightarrow z=3$

From (5), $\dfrac{1}{y}=\dfrac{1}{3} \quad \Rightarrow y=3$

From (4), $x=3$

$x=3, y=3, z=3$ **(Ans.)**

Problem 87:

$ax+by-1=0$ ....(1)

$bx+ay-\dfrac{(a+b)^2}{a^2+b^2}+1=0$ ....(2)

From (1) we get $ax+by=1$ ....(3)

From (2) we get, $bx+ay=\dfrac{2ab}{a^2+b^2}$ ....(4)

By (4) ×a, we get, $abx+a^2y=\dfrac{2a^2b}{a^2+b^2}$ ....(5)

By (1) ×b, we get, $abx + b^2 y = b$ ....(6)

Subtracting (5) and (6), we get

$a^2 y - b^2 y = \dfrac{2a^2 b}{a^2 + b^2} - b$

$\Rightarrow y(a^2 - b^2) = \dfrac{a^2 b - b^3}{a^2 + b^2}$

$\Rightarrow y = \dfrac{b}{a^2 + b^2}$ ....(7)

Substituting (7) in (3) we get,

$ax + b\left(\dfrac{b}{a^2 + b^2}\right) = 1$

$\therefore x = \dfrac{a}{a^2 + b^2}$ ....(8)

$\therefore x = \dfrac{a}{a^2 + b^2}, \; y = \dfrac{b}{a^2 + b^2}$ **(Ans.)**

Problem 88:

At 2 o'clock the hour hand is 10 minute-spaces ahead of the minute-hand.

Let, x = The number of spaces the minute-hand moves over when it coincides with the hour-hand.

Then, $(x-10)$ = The number of spaces the hour-hand moves over when it coincides with the minute-hand.

Now, the minute-hand moves 12 times as fast as the hour-hand.

$\therefore 12(x-10)$ = The number of spaces the minute-hand moves over.

$\therefore 12(x-10) = x$

$\therefore x = 10\dfrac{10}{11}$.

$\therefore$ The time is $10\dfrac{10}{11}$ minutes past 2 o'clock when both the hands coincides.

(Ans.)

Problem 89:

$$\frac{\sqrt{10}-2}{5\sqrt{3}-\sqrt{32}-\sqrt{48}+\sqrt{18}}$$

$$=\frac{\sqrt{10}-2}{5\sqrt{3}-4\sqrt{2}-4\sqrt{3}+3\sqrt{2}}$$

$$=\frac{\sqrt{10}-2}{\sqrt{3}-\sqrt{2}}$$

$$=\frac{(\sqrt{10}-2)(\sqrt{3}+\sqrt{2})}{(\sqrt{3}-\sqrt{2})(\sqrt{3}+\sqrt{2})}$$

$$=\frac{\sqrt{30}+\sqrt{20}-2\sqrt{3}-2\sqrt{2}}{(\sqrt{3}-\sqrt{2})(\sqrt{3}+\sqrt{2})}$$

$$=\frac{\sqrt{30}+\sqrt{20}-2\sqrt{3}-2\sqrt{2}}{(\sqrt{3}-\sqrt{2})(\sqrt{3}+\sqrt{2})}$$

$$=(\sqrt{10}-2)(\sqrt{3}+\sqrt{2}) \textbf{ (Ans.)}$$

Problem 90:

$$\sqrt{3+\sqrt{5}}=\sqrt{\frac{1}{2}(6+2\sqrt{5})}=\sqrt{\frac{1}{2}(\sqrt{5}+1)^2}=\frac{\sqrt{5}+1}{\sqrt{2}}$$

$$\sqrt{7-3\sqrt{5}}$$

$$=\sqrt{\frac{1}{2}(14-6\sqrt{5})}$$

$$=\sqrt{\frac{1}{2}(14-2.3.\sqrt{5})}$$

$$=\sqrt{\frac{1}{2}(9+5-2.3.\sqrt{5})}=\sqrt{\frac{1}{2}(3-\sqrt{5})^2}=\frac{3-\sqrt{5}}{\sqrt{2}}$$

$$\therefore \frac{\sqrt{3+\sqrt{5}}}{\sqrt{2}-\sqrt{7-3\sqrt{5}}}=\frac{\frac{\sqrt{5}+1}{\sqrt{2}}}{\sqrt{2}-\frac{3-\sqrt{5}}{\sqrt{2}}}=\frac{\sqrt{5}+1}{\sqrt{5}-1}=\frac{3.342+1}{3.342-1}=\frac{4.342}{2.342}=1.85 \textbf{ (Ans.)}$$

Problem 91:
$$2x^2 + 4x + 2xy + 2y + 2$$
$$= 2(x^2 + 2x + xy + y + 1)$$
$$= 2\{(x+1)^2 + y(x+1)\}$$
$$= 2(x+1)(x+y+1) \text{ (Ans.)}$$

Problem 92:
Let the speed of the car on the first day was x km/hour.

$\therefore$ The distance covered in $2\frac{1}{2}$ hours was $\frac{5x}{2}$ km.

On the second day, the distance covered by the car was $\left(\frac{5x}{2} - 1\right)$ km.

On the second day, the speed of the car was $(x+2)$ km/hour and the time taken to cover the reduced distance was 2 hours.

$$\therefore \frac{\left(\frac{5x}{2} - 1\right)}{(x+2)} = 2 \Rightarrow x = 10$$

$\therefore$ The speed of the car on the first day was 10 km/hour. **(Ans.)**

Problem 93:
Assume that the length of the train is x meter.

$\therefore$ In 25 seconds, the train covered a distance of $(x+220)$ meters while crossing the 1st bridge.

In 18 seconds, the train covered a distance of $(x+120)$ meters while crossing the 1st bridge.

Since the train runs at a constant speed,
$$\Rightarrow \frac{(x+220)}{25} = \frac{(x+120)}{18} \Rightarrow x \approx 137$$

$\therefore$ The length of the train is 137 meters.

$\therefore$ The speed of the train is $\frac{(x+220)}{25} \approx 14.3$ meters/sec i.e.,

$$\frac{14.3 \times 60 \times 60}{1000} = 51.5 \text{ km/hour}.$$

The length = 137 meters, The speed = 51.5 km/hour. **(Ans.)**

Problem 94:

Let $x = 0.\overline{7}$

$\Rightarrow 10x = 7.\overline{7}$

$\Rightarrow 10x - x = 7.\overline{7} - 0.\overline{7}$

$\Rightarrow 9x = 7$

$\Rightarrow x = \dfrac{7}{9}$

Let $y = 0.5\overline{7}$

$\Rightarrow 10y = 5.\overline{7}$

$\Rightarrow 100y = 57.\overline{7}$

$\Rightarrow 100y - 10y = 57.\overline{7} - 5.\overline{7} = 52$

$\Rightarrow 90y = 52$

$\Rightarrow y = \dfrac{52}{90}$

$\therefore 0.7 + 0.\overline{7} + 0.5\overline{7} = \dfrac{7}{10} + \dfrac{7}{9} + \dfrac{52}{90} = \dfrac{37}{18}$ **(Ans.)**

Problem 95:

Let the cost price of the cycle = $\$x$.

When the cycle was sold at 25% profit, the selling price was $x \times \dfrac{125}{100} = \$\dfrac{5x}{4}$.

In the second case, the cost price was $\dfrac{90x}{100} = \$\dfrac{9x}{10}$.

When the cycle was sold at 50% profit, the selling price was $\dfrac{9x}{10} \times \dfrac{150}{100} = \$\dfrac{27x}{20}$.

According to the question,

$\dfrac{27x}{20} - \dfrac{5x}{4} = 50 \Rightarrow x = \$500$

$\therefore$ The cost price of the cycle was $500 **(Ans.)**

Problem 96:

$$\frac{1}{(x-1)(x-2)} = \frac{1}{(x-2)} - \frac{1}{(x-1)}$$

$$\frac{1}{(x-2)(x-3)} = \frac{1}{(x-3)} - \frac{1}{(x-2)}$$

$$\frac{1}{(x-3)(x-4)} = \frac{1}{(x-4)} - \frac{1}{(x-3)}$$

$$\therefore \frac{1}{(x-2)} - \frac{1}{(x-1)} + \frac{1}{(x-3)} - \frac{1}{(x-2)} + \frac{1}{(x-4)} - \frac{1}{(x-3)} = \frac{1}{6}$$

$$\Rightarrow \frac{1}{(x-4)} - \frac{1}{(x-1)} = \frac{1}{6}$$

$$\Rightarrow x^2 - 5x - 14 = 0$$

$$\Rightarrow (x+2)(x-7) = 0$$

$\therefore x = -2$ or $x = 7$ **(Ans.)**

Problem 97:

$x^2 - x = 1482$

$$\Rightarrow x^2 - x + \left(\frac{1}{2}\right)^2 = 1482 + \left(\frac{1}{2}\right)^2$$

$$\Rightarrow \left(x - \frac{1}{2}\right)^2 = \frac{5929}{4}$$

$$\Rightarrow \left(x - \frac{1}{2}\right)^2 = \left(\frac{77}{2}\right)^2$$

$$\Rightarrow \left(x - \frac{1}{2}\right)^2 - \left(\frac{77}{2}\right)^2 = 0$$

$$\Rightarrow \left(x - \frac{1}{2} + \frac{77}{2}\right)\left(x - \frac{1}{2} - \frac{77}{2}\right) = 0$$

$$\Rightarrow (x+38)(x-39) = 0$$

$\therefore x = -38$ or $x = 39$ **(Ans.)**

Problem 98:

$(x-2)(x-3) = \dfrac{34}{33^2}$

Suppose $a = 33$

$\Rightarrow (x-2)(x-3) = \dfrac{34}{a^2}$

$\Rightarrow a^2(x-2)(x-3) = 33+1$

$\Rightarrow (ax-2a)(ax-3a) = a+1$

$\Rightarrow (ax-2a)(ax-3a) - a - 1 = 0$

$\Rightarrow (ax-2a)(ax-3a) + (ax-3a) - (ax-2a) - 1 = 0$

$\Rightarrow (ax-3a)(ax-2a+1) - (ax-2a+1) = 0$

$\Rightarrow (ax-2a+1)(ax-3a-1) = 0$

$\Rightarrow (ax-2a+1) = 0$ or $(ax-3a-1) = 0$

$(ax-2a+1) = 0$

$\Rightarrow x = \dfrac{2a-1}{a} = \dfrac{65}{33}$

$(ax-3a-1) = 0$

$\Rightarrow x = \dfrac{3a+1}{a} = \dfrac{100}{33}$

$\therefore x = \dfrac{65}{33}$ or $x = \dfrac{100}{33}$ **(Ans.)**

Problem 99:

$4xy = (x+y)^2 - (x-y)^2$

The value of $4xy$ is greatest when $(x-y)^2$ is smallest.

However, $(x-y)^2 \geq 0$; so minimum value is zero.

$\therefore 4xy = (x+y)^2 = 100 \Rightarrow xy = 25$ **(Ans.)**

Problem 100:

$(x+y+z)^2 \geq 0$

$\Rightarrow x^2 + y^2 + z^2 + 2(xy + yz + zx) \geq 0$

$\Rightarrow 1 + 2(xy + yz + zx) \geq 0$

$\Rightarrow 2(xy + yz + zx) \geq -1$

$\Rightarrow (xy + yz + zx) \geq -\dfrac{1}{2}$ ...(1)

Again,

$(x-y)^2 + (y-z)^2 + (z-x)^2 \geq 0$

$\Rightarrow x^2 + y^2 + z^2 - xy - yz - zx \geq 0$

$\Rightarrow 1 - xy - yz - zx \geq 0$

$\Rightarrow 1 \geq xy - yz - zx$ ...(2)

From (1) and (2)

$-\dfrac{1}{2} \leq xy - yz - zx \leq 1$ **(Ans.)**

ABOUT THE AUTHOR

Sanjay Jamindar, a Wireless Telecommunication Engineer and Consultant by profession, has to his credit many a technical contribution to various clientele in the field of Telecommunication within the Information Technology industry. During his professional career, he published many technical papers in national and international conferences.

Sanjay has a Bachelor's degree in Physics (Hons.) and Bachelor's in Engineering (Electronics & Communication) from Calcutta University, Kolkata, India. He completed his Master's (M.Tech) in Laser Technology and Optical Communication from Indian Institute Of Technology (IIT), Kanpur, India.

He received National Scholarships during Schooling and Bachelor's degree (Physics) Examinations. He also received the DAAD (German Govt.)-Scholarship, to pursue Master's thesis work in Germany (Technical University, Berlin).

Sanjay's primary interest lies in research in the domain of Telecommunication. He also writes technical books including various educational books that can help students in strengthening their knowledge, in the domain of Mathematics.

www.ingramcontent.com/pod-product-compliance
Lightning Source LLC
Chambersburg PA
CBHW081123180526
45170CB00008B/2979